Starry Nights

Austin P. Torney

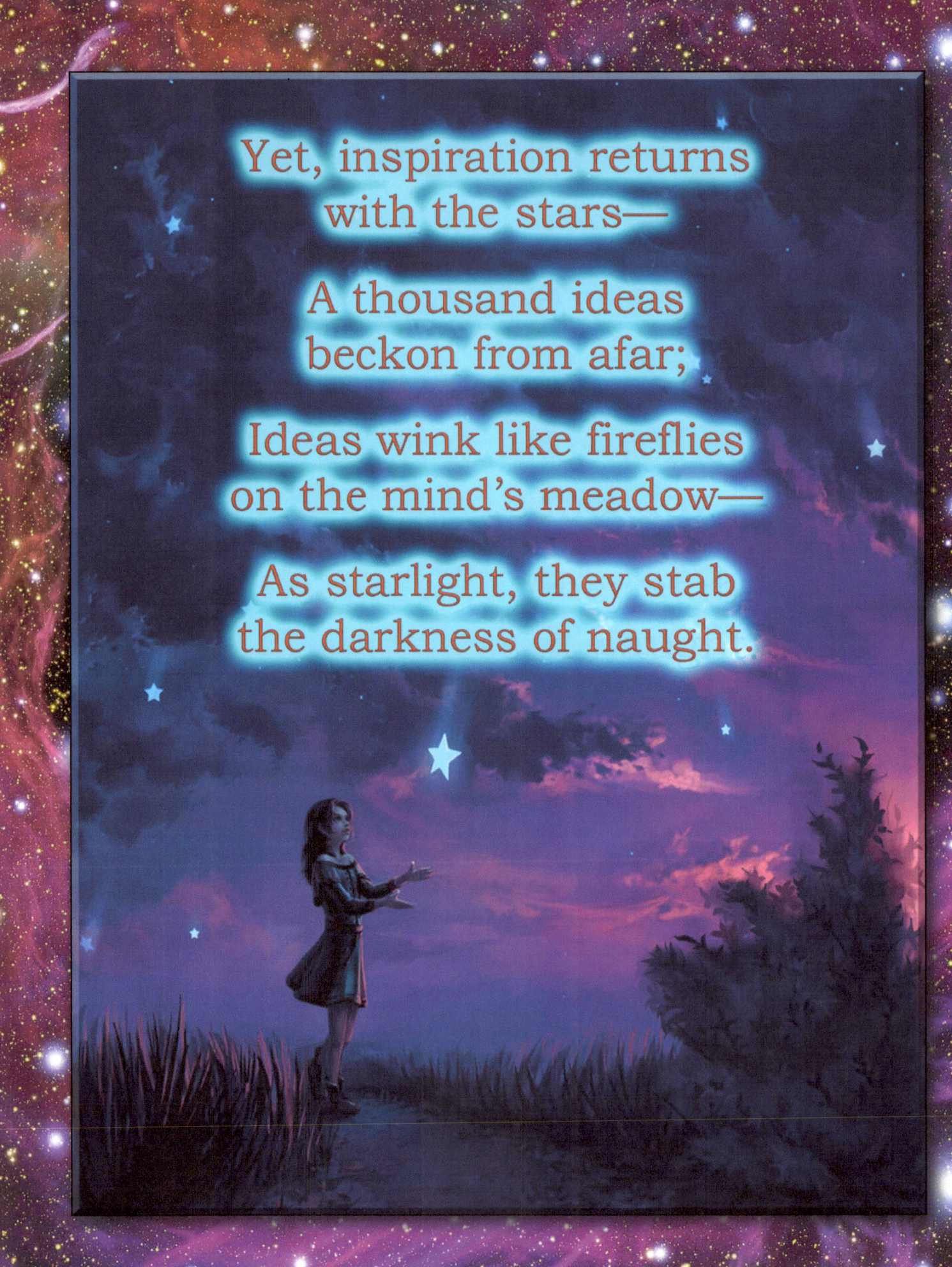

Yet, inspiration returns with the stars—

A thousand ideas beckon from afar;

Ideas wink like fireflies on the mind's meadow—

As starlight, they stab the darkness of naught.

The stars' light is
the origin of our being,

The source of our matter,
energy—everything;

Permanent, reassuring,
and unquenchable,

It's our radiant soul,
our self-winding mainspring.

Stars generate
the lower elements;

Supernovae generate
the higher ones.

Atoms form the molecules
that lead to

Life's complexity,
from simplicity.

Purgatory's on Venus,
where sulfurs rain.

Hell's found in the sun's heart,
oh, hot burning pain!

Of Heaven's site,
no one has any idea—

It's the world's best kept secret:
Earth's its name!

Earth's a garden, an oasis in space,
A world of boundless beauty and grace.

One could search
the heavens for such in vain,

Finding no equal,
anytime or anyplace.

We are life's
eternal creative smile,

Beaming as
the universal epistyle.

In us the cosmos
has come alive;

Thus we borrow life
from Death for awhile.